QuickStudy.

for

Biology

Boca Raton, Florida

DISCLAIMER:

This QuickStudy® Booklet is an outline only, and as such, cannot include every aspect of this subject. Use it as a supplement for course work and textbooks. BarCharts, Inc., its writers and editors are not responsible or liable for the use or misuse of the information contained in this booklet.

©2006 BarCharts, Inc.

Animal and Plant Cell Images (pages 13 & 14) provided by Molecular Expressions (*microscopy.fsu.edu*)

ISBN 13: 9781423202561

ISBN 10: 1423202562

BarCharts® and QuickStudy® are registered trademarks of BarCharts, Inc.

Author: Randy Brooks, Ph.D.

Publisher:

   BarCharts, Inc.

   6000 Park of Commerce Boulevard, Suite D

   Boca Raton, FL 33487

   www.quickstudy.com

Printed in Thailand

Contents

Study Hints

NOTE TO STUDENT:

Use this QuickStudy® booklet to make the most of your studying time.

QuickStudy® examples offer detailed explanations; refer to them often to avoid problems.

Examples:
- **Phagocytosis** = "Cell Eating"
- **Pinocytosis** = "Cell Drinking"

QuickStudy® notes provide need-to-know information; read them carefully to better understand key concepts.

NOTES

Evolution is the concept that all organisms are related to each other by common ancestry; which is the unifying theme in biology.

Take your learning to the next level with Quick Study®!

Basic Concepts

NOTES
Bio = Life
Logy = Study of

Biological Science: The Study of Life

The Scientific Method
■ How scientists study biology
- ◆ Observe phenomena
- ◆ Formulate testable and falsifiable (in case they are wrong) hypotheses
- ◆ Test hypotheses
- ◆ Collect data
- ◆ Analyze statistically (if necessary)

What Is Life?
■ Characteristics:
- ◆ metabolism
- ◆ reproduction
- ◆ growth
- ◆ movement
- ◆ responsiveness
- ◆ complex organization

2 Evolution: Overview

NOTES
Evolution is the concept that all organisms are related to each other by common ancestry, which is the unifying theme in biology.

■ **Natural Selection:** A mechanism for the occurrence of evolution.
◆ **Survival** of those offspring best adapted to the conditions in which they live:
  • Individuals produce sexually many more offspring than could possibly survive.
  • These offspring are not identical (in most situations), but show variations based on genetic differences.
  • Essentially, those individuals with variations that allow them to survive (i.e., **adaptations**) to the age of reproduction can pass their genes on to the next generation.
  • Thus, nature is selecting offspring and shaping the evolution of species.

■ **Charles Darwin** and **Alfred Wallace**, 19th century biologists, formulated the concept of **natural selection.**

Organismal Evolution

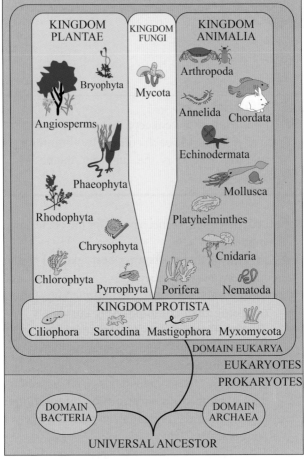

■ **Artificial Selection:** Human selects traits in offspring.

Examples:
Uses for artificial selection include pets and agricultural crops.

Domesticated Animals

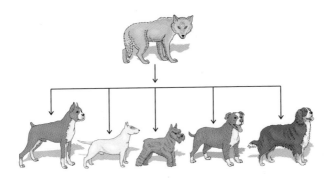

Cytology

NOTES
Cytology is the study of cells.

Cell Theory
All living things are composed of cells and come from cells.

■ **Cell Size:** Small to maximize surface area to volume ratio for regulating internal cell environment

■ **Cell (Plasma) Membrane:** Composed of fluid-like phospholipid bilayer, proteins, cholesterol and glycoproteins

■ **Cell Wall:** Outside of cell membrane in some organisms; composed of carbohydrate (e.g., **cellulose** or **chitin**) or carbohydrate derivative (e.g., **peptidoglycan**)

Cell (Plasma) Membrane

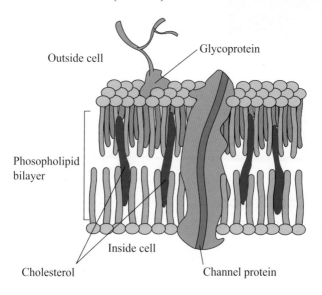

Outside cell

Glycoprotein

Phosopholipid
bilayer

Inside cell

Cholesterol

Channel protein

■ **Cytoplasm:** Material outside nucleus
   ◆ Site for metabolic activity
   ◆ **Cytosol:** Solutions with dissolved substances such as glucose, CO_2, O_2, etc.
   ◆ **Organelles:** Membrane-bound subunits of cells each specialized functions

■ **Cytoskeleton:** Supportive and metabolic structure composed of microtubules, microfilaments and intermediate filaments

NOTES
Prokaryotic = No membrane-bound organelles and no nucleus
Eukaryotic = Membrane-bound organelles plus nucleus

Cytoskeleton

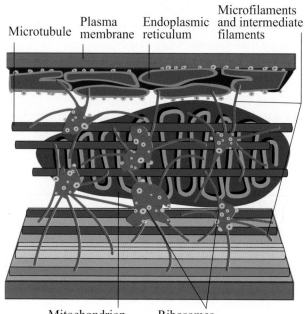

Microtubule   Plasma membrane   Endoplasmic reticulum   Microfilaments and intermediate filaments

Mitochondrion   Ribosomes

Prokaryotic Cells

Simpler cellular organization with no nucleus or other membrane-bound organelles

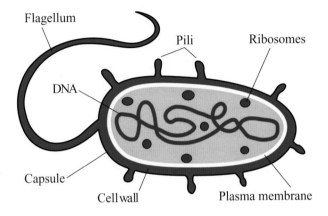

Eukaryotic Cells

Complex cellular organization

■ **Membrane:** Bound organelles including the following:

◆ **Nucleus:** DNA/chromosomes, control cellular activities via genes

◆ **Nucleolus:** Located within nucleus, site for ribosome synthesis

◆ **Rough endoplasmic reticulum:** With ribosomes, involved in protein synthesis

◆ **Smooth endoplasmic reticulum:** Without ribosomes, involved primarily in lipid synthesis

◆ **Golgi apparatus:** Packaging center for molecules; carbohydrate synthesis

◆ **Lysosome:** Contains hydrolytic enzymes for intracellular digestion

- ◆ **Peroxisome:** Involved in hydrogen peroxide synthesis and degradation
- ◆ **Chloroplast:** Site of photosynthesis
- ◆ **Chromoplast:** Non-green pigments
- ◆ **Leukoplast:** Stores starch
- ◆ **Mitochondrion:** ATP production
- ◆ **Vacuole:** General storage and space-filling structure

Animal Cell

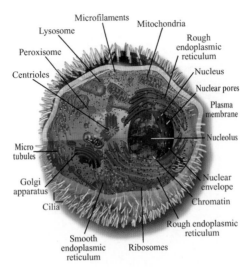

Plant Cell

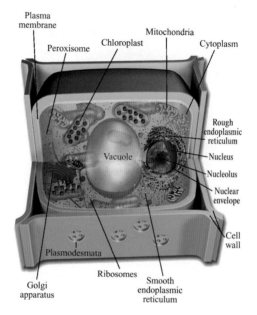

Plasma membrane
Peroxisome
Chloroplast
Mitochondria
Cytoplasm
Rough endoplasmic reticulum
Nucleus
Nucleolus
Nuclear envelope
Vacuole
Cell wall
Plasmodesmata
Ribosomes
Smooth endoplasmic reticulum
Golgi apparatus

NOTES
Animals and plants have **eukaryotic** cells.

Energy & Life

Our Sun

Organisms must use the sun's energy (directly or indirectly) to become and remain in an organized state.

Metabolism

Series of chemical reactions involved in storing (**anabolism**) or releasing (**catabolism**) energy.

Enzymes

Biological Catalysts: Facilitate metabolic chemical reactions by speeding up rates and lowering heat requirements.

Adenosine Triphosphate (ATP)

A high-energy molecule: Energy stored in ATP is released by breaking phosphate-to-phosphate bonds and creating adenosine diphosphate (ADP) or adenosine monophosphate (AMP).

■ ATP is recycled by adding back phosphate groups using energy from the sun.

Enzyme Kinetics

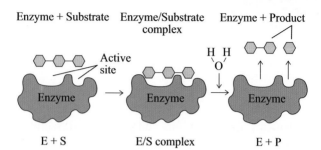

Enzyme + Substrate   Enzyme/Substrate complex   Enzyme + Product

E + S   E/S complex   E + P

NOTES

Enzymes are essential to metabolic processes.

Energy & ATP

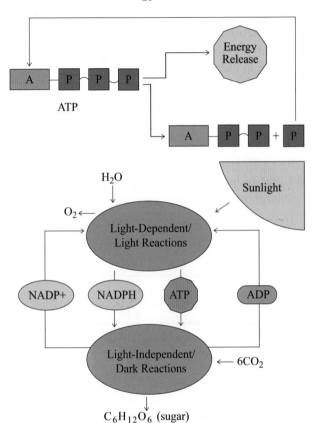

Photosynthesis

Sunlight or radiant energy is **captured** by **chlorophyll** and **carotenoid photopigments**
■ Found in **cytoplasm** in *prokaryotes*
■ Found in **chloroplasts** in *eukaryotes*

Two Main Processes

■ **Light-dependent reactions (Light Reactions):** The captured light energy is transferred to electrons that come from H_2O. O_2 is a by-product.

■ **Light-independent reactions (Dark Reactions):** Energized electrons are transferred to CO_2 (reduction reactions) to form glucose (in the Calvin-Benson cycle).

Examples:
Sugar is created by a light-independent reaction.
[*See* **Energy & ATP,** page 17.]

Cell Respiration

Highly energized electrons stored temporarily in glucose are removed (oxidation reactions) in a step-wise fashion to maximize energy capture at each step:

■ **Glycolysis:** Anaerobic process in cytoplasm in which glucose, a six-carbon compound, is oxidized to two **pyruvates**, which are both three-carbon chains.

■ **Krebs Cycle:** Aerobic process that oxidizes pyruvates to CO_2.

■ **Chemiosmotic Phosphorylation:** The energized electrons released during the previous steps are used to concentrate hydrogen ions in one area (of the cell membrane in prokaryotes; of the mitochondrion in eukaryotes) to create a chemical gradient between positively and negatively charged ions (i.e., a battery).

NOTES
The potential energy resulting from this osmotic gradient is used to resynthesize ATP from ADP and AMP. After electrons have been used, they must be transferred to O_2.

Cell Transport

Passive Transport

■ Relies on thermal energy of matter; the cell does not do work.

■ There are four categories:

◆ **Diffusion:** Movement from an area of high to low concentration.

◆ **Facilitated Diffusion:** A permease, or membrane enzyme, carries substance

◆ **Osmosis:** Diffusion across a semi-permeable membrane

◆ **Bulk Flow:** Mass movements of fluids affected by pressure and solutes

Osmosis

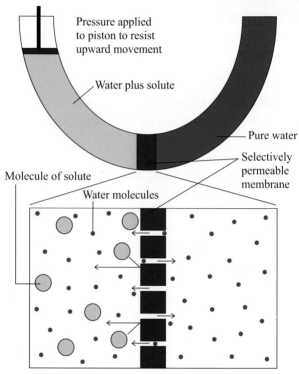

Pressure applied to piston to resist upward movement

Water plus solute

Pure water

Selectively permeable membrane

Molecule of solute

Water molecules

⟵ Net movement of water molecules

Membrane Pump – ATP Required

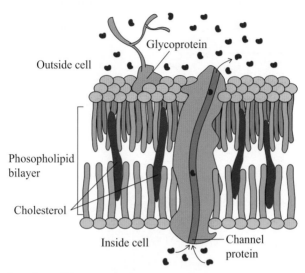

Active Transport

■ Relies on the cell providing energy supply, there are three categories:

◆ **Membrane Pumps:** Permease used to move substance, usually in the opposite direction of diffusion

◆ **Endocytosis:** Materials are brought into cell via:
  • **Phagocytosis:** Solids
  • **Pinocytosis:** Liquids

◆ **Exocytosis:** Expel materials from cell

Endocytosis

Phagocytosis

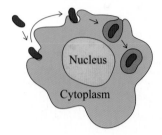

Pinocytosis

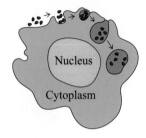

Exocytosis

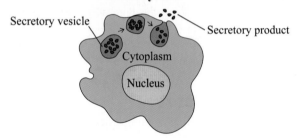

Examples:
- **Phagocytosis** = "Cell Eating"
- **Pinocytosis** = "Cell Drinking"

6 Cell Reproduction

NOTES
Cells reproduce in two steps:
Mitosis and **Cytokinesis**

■ **Mitosis:** Division of nuclear material
■ **Cytokinesis:** Division of remaining cellular contents of the cytoplasm

Cell Cycle
■ Cells go through four stages:
  ◆ **G1:** Active growth and metabolism
  ◆ **S:** DNA synthesis and duplication
  ◆ **G2:** Synthesis of molecules in preparation for cell division
  ◆ **Mitosis & Cytokinesis**

Examples:

Stages **G1**, **S**, & **G2** above are collectively referred to as Interphase.

Interphase chromosomes are referred to as chromatin, a diffuse, loosely scattered arrangement of chromosomes.

Mitotic chromosomes in the mitosis/cytokinesis stage are highly condensed and coiled, and thus distinct.

Mitosis: Four Mitotic Stages

■ **Prophase:**
   ◆ **Chromosomes** condense and organize
   ◆ **Nuclear** membrane and nucleoli disappear
   ◆ **Spindle** apparatus assembled and attached to **centromeres** of duplicated chromosomes

■ **Metaphase:**
   ◆ **Spindles** line up duplicated chromosomes along equator of cell.
   ◆ One spindle to each half or **chromatid** of duplicated chromosome

■ **Anaphase:**
   ◆ Centromere of each duplicated chromosome is separated and
   ◆ **Paired chromatids** are pulled apart

■ **Telophase:**
   ◆ Chromosomes uncoil
   ◆ Nucleoli reappear
   ◆ Cytokinesis occurs
   ◆ Two genetically **identical daughter** cells are produced

Cell Cycle

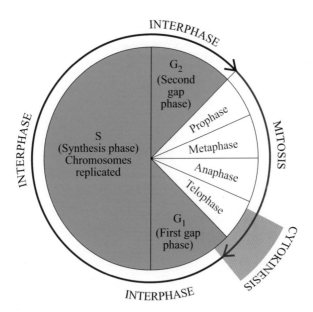

Mitosis

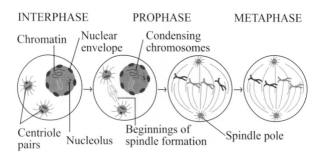

INTERPHASE   PROPHASE   METAPHASE

Chromatin — Nuclear envelope   Condensing chromosomes

Centriole pairs   Nucleolus   Beginnings of spindle formation   Spindle pole

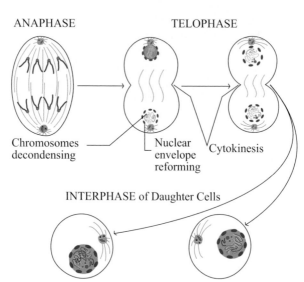

ANAPHASE   TELOPHASE

Chromosomes decondensing   Nuclear envelope reforming   Cytokinesis

INTERPHASE of Daughter Cells

Two new cells are genetically identical (i.e., clones)

7 Organismal Reproduction & Meiosis

Sexual Processes
■ Sexual Reproduction:
- ◆ Fusion of Genetic material (gametes) from two parental organisms
- ◆ To ensure the proper numbers in the **zygote** (fertilized egg), each gamete must have half or **haploid** (N) of the original **diploid** (2N) amount of DNA

Examples:
Sperm = N
Skin Cell = 2N

- ◆ **Meiosis:** Reduces the **chromosome** number by half and results in new genetic combinations in the gametes
  - • 2 distinct stages
  - • Preceded by interphase
  - • Many meiotic events similar to mitosis.
  - • Differences are noted below.

NOTES
Meiosis I & II have the same stage names
- • Prophase
- • Metaphase
- • Anaphase
- • Telophase

■ **Meiosis I**
  ◆ **Prophase I:** Chromosomes condense and organize and matched or homologous chromosomes (one maternal and one paternal in each pair) are physically paired. Segments of chromatids can cross over within each chromosome pair
  ◆ **Metaphase I:** Homologues line up at equator
  ◆ **Anaphase I:** Homologues separated into two groups, with each group having a mixture of maternal and paternal chromosomes
  ◆ **Telophase I:** New haploid nuclei forming for two new daughter cells
  ◆ **Interkinesis:** No replication of DNA occurs because each chromosome is still duplicated and consists of two chromatids (although crossing over results in some chromatids with maternal and paternal segments).

■ **Meiosis II**
  ◆ **Prophase II:** Chromosomes condense.
  ◆ **Metaphase II:** Chromosomes line up at equator.
  ◆ **Anaphase II:** Chromatids of each chromosome are separated.
  ◆ **Telophase II:** Each daughter cell from meiosis I will form two more cells for a total of four cells.

Faunal/Floral Gametogenesis
■ In animals, meiosis occurs in germinal tissues and is called **spermatogenesis** in **males** and **öögenesis** in **females.** Each results in a gamete.
■ In plants the process is similar except that mitotic divisions may follow meiosis to produce gametes.

Meiosis I

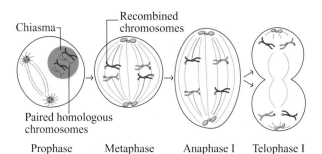

Chiasma
Recombined chromosomes
Paired homologous chromosomes

Prophase Metaphase Anaphase I Telophase I

Meiosis II

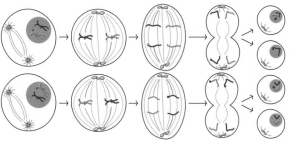

Prophase II Metaphase II Anaphase II Telophase II Four daughter cells*

*Four new cells are genetically unique and haploid.

Cross Over

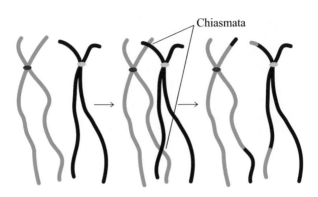

Chiasmata

Gametogenesis

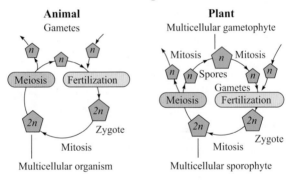

Animal
Gametes

Plant
Multicellular gametophyte

8 Genetics & Mendel

NOTES
Genetics is the study of traits and their inheritance.

Introduction

■ 19th century biologists believed that traits blended. If blending occurred, things would become more similar, not different. **Darwin** and **Wallace** stated that variations or differences in offspring were necessary for natural selection to occur.

■ **Gregor Mendel** provided the most plausible hypothesis for genetics.

◆ **Mendelian Genetics:**
• Two laws were developed by using statistics to analyze results of crosses involving distinguishing traits of garden peas.

Law of Segregation of Alternate Factors

Developed by Mendel using single-trait crosses

■ Single-trait crossbreeding:

◆ Two **true-breeding** (those that consistently yield the same form when crossed with each other) parents (P) but different strains were crossed (e.g., round versus wrinkled seed)

◆ The **offspring** (F_1) from this cross all showed only one trait (e.g., round seed) and this was called the **dominant trait.** The traits from the parents did not blend.

◆ The F_1 individuals were crossed with each other to produce F_2 individuals.

◆ Three-fourths of the F_2 expressed the **dominant trait.**

◆ One-fourth expressed the trait of the other P1 parent (e.g., wrinkled seed) which had not been expressed in the F_1 generation and was thus recessive.

■ Mendel's crosses for single traits can be summarized as follows:

Mendel's First Law:
Segregation of Alternate Factors

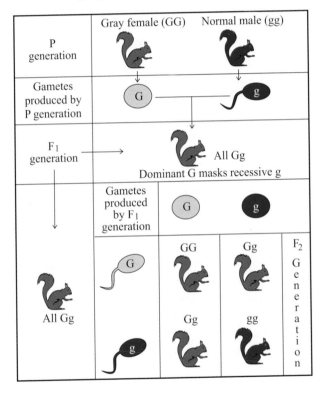

■ Mendel's First Conclusions:

Discrete factors (now known as genes) were responsible for the traits and these factors were paired, separated (which occurs during meiosis) and recombined (during fertilization).

◆ Alternate forms of factors or genes exist called alleles.

- The F_1 individuals had two alleles, their genotype consisted of a dominant and recessive allele (e.g., Rr with R for round and r for wrinkled seed).
- Thus, the F_1's were hybrids.
- Their **phenotype** was similar to only one of original parents (e.g., round seed).

Mendel Updated

■ Genes are found on chromosomes, and thus multiple traits assort independently as long as they are located on different chromosomes. Mendel studied traits in peas that were each on separate chromosomes. Genes on the same chromosome are linked and thus will not normally assort independently.

■ Interactions between alleles:

◆ **Complete dominance:** One allele dominates another allele

◆ **Incomplete dominance:** Neither allele is expressed fully

◆ **Codominance:** Both alleles are expressed fully

◆ **Multiple alleles:** More than two alleles for a gene are found within a population

◆ **Epistasis:** One gene alters the affect of another gene

◆ **Polygenic inheritance:** Many genes contribute to a phenotype

◆ **Pleiotropy:** One gene can effect several phenotypes.

◆ **Environmental influences:** Where the genotype and environment interact to form a phenotype

Law of Independent Assortment

Developed by Mendel using multiple-trait crosses.

■ Two true-breeding parents of different strains for two traits were crossed. The F_1's were then crossed, producing F_2 individuals.

■ The results of crosses involving two traits can be summarized in the diagram on page 38.

■ Mendel concluded statistically that these results occurred because alleles for one trait or gene did not affect the inheritance of alleles for another trait.

Mendel's Second Law: Independent Assortment

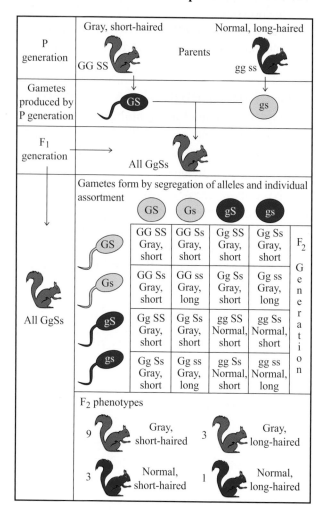

Chromosomes & Sex Determination

■ In many animals special chromosomes determine sex, the remaining chromosomes are autosomes.

■ In humans there are 44 autosomes and two sex chromosomes: X and Y in males, X and X in females.

Sex Determination

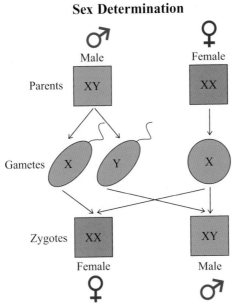

Sex-Linked Traits

In humans, the Y chromosome contains the determinant for maleness, the X contains many genes. If a male gets a recessive (or dominant) allele on the X chromosome from his mother, he will express the trait. Therefore, males are frequently afflicted with X-linked disorders.

Examples:
In genetics
$$XY = male \qquad XX = female$$

9 Molecular Genetics

Genes, DNA & Nucleic Acid

■ **Gene Functions**
   ◆ To be preserved and transmitted
   ◆ To control various biological functions through the production of proteins (i.e., large, complex sequences of amino acids) and RNA

■ **Gene Structure** – two types of nucleic acids:
   ◆ Deoxyribonucleic acid (DNA)
   ◆ Ribonucleic acid (RNA)

■ **Nucleotides** – the components of nucleic acids

Nucleotides

O
O-
P
-O
O
Phosphate
CH2
Sugar
H
H
H
H
OH
H
H
N
H
N
H
N-H
N
N
H
Nitrogenous
base

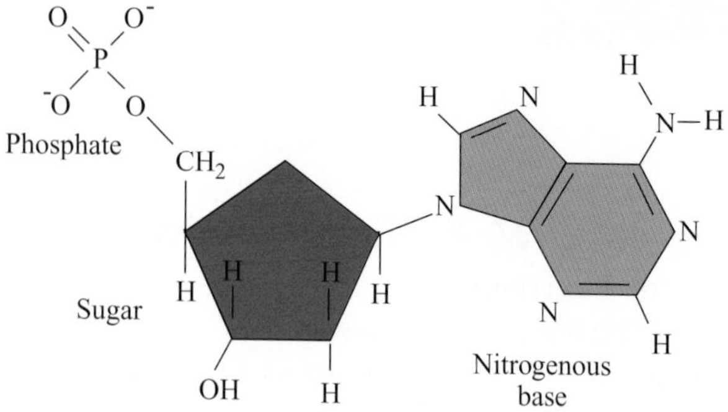

■ **Three Subunits of Nucleotides**
  ◆ **Sugar** (deoxyribose in DNA; ribose in RNA)
  ◆ **Phosphate**
  ◆ **Nitrogenous Base** (5 possible bases)
    • In DNA, the nucleic acid of chromosomes, four nitrogenous bases are found:
      > Adenine (A),
      > Guanine (G),
      > Cytosine (C),
      > Thymine (T)

    • RNA consists of similar bases, except uracil (U) replaces thymine (T)

DNA Double Helix

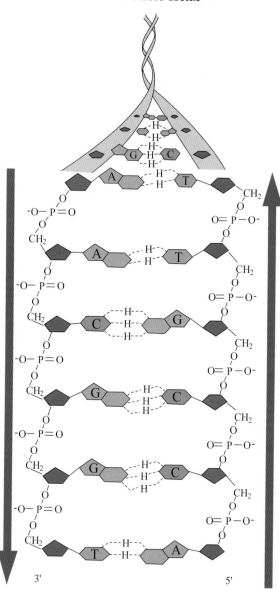

• **DNA** is a **double helix** molecule.

NOTES
Similar to a spiral staircase or twisted ladder, with the sides formed by repeating sugar-phosphate groups from each nucleotide, and the horizontal portions (i.e., steps) formed by hydrogen bonds involving A with T or C with G.

• Hereditary information: (i.e., genes) found along the linear sequence of nucleotides in the DNA molecule.

The Central Dogma
■ Replication
◆ DNA is copied from other DNA by unzipping the helix and pairing new nucleotides with the proper bases (i.e., A with T and C with G) on each separated side of the original DNA.

■ Transcription
◆ **Messenger (m)RNA** is copied from DNA by unzipping a portion of the DNA helix that corresponds to a gene.

◆ Only one side of the DNA will be transcribed and nucleotides with the proper bases (A with U and C with G) will be sequenced to build pre-(m)RNA.

◆ Sequences of nucleotides called **introns** are removed, and the remaining segments (called **exons**) are *spliced* together.

◆ The mature mRNA leaves the nucleus to be transcribed by the ribosomes.

■ Translation

◆ Proteins are synthesized from (m)RNA by ribosomes which are composed of ribosomal (r)RNA and proteins) that read from a universal triplet code (i.e., **codons**).

◆ The ribosomes instruct transfer (t)RNA's to bring in specific amino acids in the sequence dictated by the mRNA, which in turn was built based on the sequence of nucleotides in the original gene portion of the DNA.

Examples:
- **(m)RNA** = messenger
- **(r)RNA** = ribosomal
- **(t)RNA** = transfer

Mutations

Any random, permanent change in the DNA molecule. Many are harmful, some have no effect, and a few actually benefit the organism. Nature selects those mutations that are beneficial or adaptive in organisms to help shape the course of evolution.

RNA Synthesis/Transcription

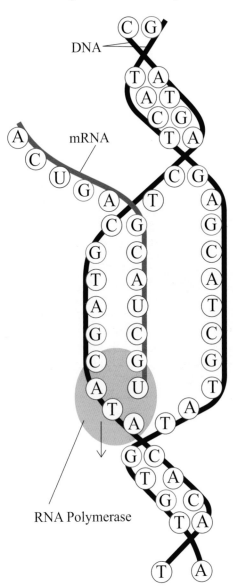

Translation/Protein Synthesis

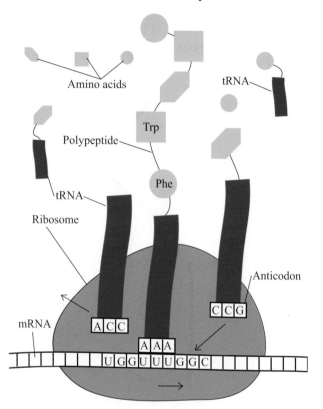

10 Population Genetics

> **NOTES**
> **Population genetics** is the study of genes in populations versus individuals. Populations evolve just as species do. To understand changes in gene pools as populations evolve, an understanding of non-evolving populations is necessary.

Population vs. Individual
- **Genotype:** Genetic composition of an individual
- **Gene Pool:** Genetic composition of a population of individuals (i. e., all alleles for all genes in a population)
- **Evolution:** Changes in gene pools over time.

The Hardy-Weinberg Law
- Both **allelic frequencies** and **genotypic ratios** (i.e., gene pools) remain constant from generation to generation in sexually producing populations, if the following conditions of equilibrium exist:
  - ◆ Mutations do not occur.
  - ◆ No net movement of individuals out of or into a population occurs.

◆ All offspring produced have the same chances for survival, and mating is random (i.e., no natural selection occurs)
◆ The population is large so that chance would not alter frequencies of alleles.
■ Algebraic equivalent of the Hardy-Weinberg Law:
◆ $p^2 + 2pq + q^2 = 1$, **where**:
  • p = frequency of dominant allele
  • q = frequency of recessive allele
  • p^2 = AA genotype
  • 2pq = Aa genotype
  • q^2 = aa genotype

Examples:

◆ If in a group of six individuals there are nine dominant (A) alleles and three recessive (a) alleles, then p = 9/12 or 0.75 and q = 3/12 or 0.25. A total of 12 gametes will be produced, nine of which will have the dominant allele and three with the recessive allele.

◆ The algebraic equation above can be used to predict the ratios of the three possible genotypes as a result of fertilizations.
  • Frequency of AA genotypes is p^2 or $(0.75)2$ = 0.56.
  • Frequency of Aa genotypes is 2pq or 2(0.75)(0.25) = 0.38.
  • Frequency of aa genotypes is q^2 or $(0.25)2$ = 0.06.

◆ The frequencies of dominant and recessive alleles is still the same — the specific alleles have been redistributed.

Hardy-Weinberg & Natural Populations

■ Few (if any) populations are in equilibrium. Therefore, changes in allele frequencies and thus gene pools do occur in natural populations.

■ The Hardy-Weinberg Law helps to identify the mechanisms of these evolutionary changes by predicting that one or more of the four conditions required are not met. That is:

◆ Mutations occur.

◆ Individuals leave and enter populations.

◆ Nonrandom mating and natural selection occur.

◆ Small populations exist.

Allele Frequency Changes

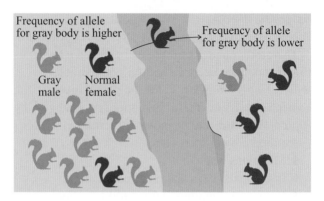

Evolution: A Review

NOTES

Remember that evolution is a science based on the concept that all organisms are related by common ancestry. It is a fundamental paradigm of biology

Summary of Definitions

■ **Natural Selection:** The mechanism for how evolution occurs

◆ Species have developed high potential for rapid **reproduction.**

◆ Population sizes eventually level off and remain fairly constant over time.

◆ **Competition** for reproduction and survival of offspring occurs.

◆ **Variations** (from random **mutations** and shuffling of genes via meiosis) exist in behavior, physiology, structure, etc.

◆ Nature selects individuals (i.e., the **fittest**—or just fortunate) for survival and reproduction to pass these favorable characteristics **(adaptations)** via their genes to their offspring.

◆ Over time, natural selection "can" lead to genetic changes in populations (i.e., evolution).

◆ **Microevolution:** Small-scale changes

◆ **Macroevolution:** Larger-scale changes—can lead to evolution of new species and groups.

12 | Evidence for Evolution I

NOTES

Remember the **cell theory:**

◆ The cell is the basic unit of life.

◆ Every life form, from bacteria to humans, is made of/comes from this basic structure.

Cellular & Molecular Perspective

■ **Organic Molecules**

◆ All life (i.e., 99 percent) consists of carbon, hydrogen, oxygen, nitrogen, phosphorus and sulfur.

◆ Evolutionary relatedness explains organisms' common usage of a small subset of 90 available elements.

■ **DNA**

◆ Genetic informational molecule in every organism, including viruses (which appear to be molecular fragments of DNA/RNA capable of "living" in host cells)

◆ DNA "language" **(Genetic Code)** is essentially universal, but slightly different dialects exist in some single-celled organisms and in some mitochondrial/chloroplast **genomes**.

◆ A common genetic language allows for such phenomena as the insertion of human genes into bacteria, which can then produce "human" proteins [*see* chapter 16, **Molecular Biotechnology**].

■ **ATP** (Adenosine triphosphate) is the primary energy currency molecule used by every organism.

DNA Double Helix

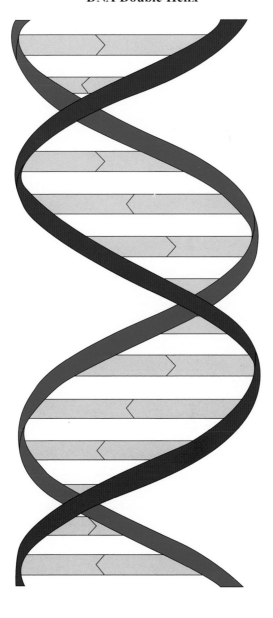

13 Evidence for Evolution II

NOTES
Natural selection (*survival of the fittest*) can be controlled and manipulated by man.

Artificial Selection

■ **Human-controlled breeding** of species strongly supports the idea that, over time, nature could also influence changes in populations.

◆ Humans have selected for traits to increase the attractiveness (to us) of the offspring (e.g., "cute" dogs, chickens that produce many eggs, wheat that yields numerous, plump grains).

◆ Domesticated species often do poorly in the wild, as traits (i.e., variations) selected by humans would not necessarily be advantageous in nature.

Artificial Selection for Crop Production

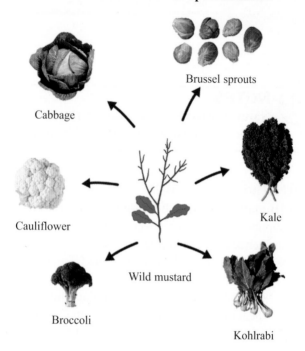

Cabbage

Brussel sprouts

Cauliflower

Kale

Wild mustard

Broccoli

Kohlrabi

Biogeography

■ Geographic distribution of species can show how organisms are related.

■ Flightless birds, such as African ostriches, Australian emus, and South American rheas are found (naturally) only in the southern hemisphere; on separate continents.

■ Either flightlessness in these birds evolved independently three times (possible, but improbable) or they arose from a common, flightless ancestor.

■ If the latter explanation is correct, and they could not fly, how then could they get to these disparate southern continents while being excluded from the northern hemisphere?

Plate Tectonics & Continental Drift

■ Geological evidence indicates the continents were once one large land mass that subsequently broke up into pieces **(plate tectonics)** that moved **(continental drift)** first into northern and southern portions, and later into the present–day continents.

◆ This geological concept also explains why **marsupial mammals** (e.g., kangaroo) developed only on Australia, as this continent was geographically isolated from areas where **placental mammals** evolved.

Fossils

■ Fossils are preserved remnants of dead organisms.

■ Darwin termed evolution: **"descent with modification."**

■ Although the fossil record has gaps (some structures/organisms do not fossilize well), fossils provide valuable information about evolutionary changes or modifications in organisms, which have taken place over many generations, including **transitional** forms:
  ◆ horses with toes
  ◆ whales with hind limbs
  ◆ ferns with seeds

■ Estimating the age of fossils involves looking at their physical positions in sedimentary rocks **(relative dating)** and radiometric isotope techniques **(absolute dating).**

■ **Molecular clocks** look at changes in portions of genomes of organisms; also used to help determine the age of evolutionary events.

Homologies

■ **Anatomical similarities** of related life forms are called homologies.

■ They provide strong evolutionary evidence of relatedness.

Examples:
• **Forelimbs** of **vertebrates** are composed of the same basic bones in disparate groups, but differ based on adaptations necessary for the specific environmental needs (i.e., walking, swimming, flying).

◆ Vestigial structures

- Those present are usually in a rudimentary, non-functional form
- Show anatomically-related structures that are likely to disappear completely in future generations.

Examples:
- The vestiges of **pelvic bones** within the body in some modern-day **baleen whales**.

Homologous Forelimb Bones: Evidence for Vertebrate Evolution

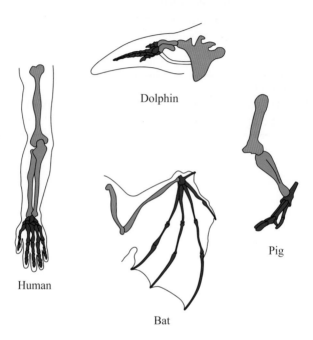

Dolphin

Human

Bat

Pig

Variations in Life

▣ In England, the peppered moth shifted from predominantly light coloring to dark when air pollution darkened the trees on which they live.

▣ Predators can easily spot moths that contrast with their background, limiting the abundance of these types of moths in the population.

▣ Subsequent air quality measures have lightened trees, and light-colored moths are again the predominant form.

▣ Additional examples of selection observed in living organisms involve increasing **drug resistance:**

Examples:
◆ bacteria resistant to antibiotics
◆ insect resistant to insecticides
◆ viruses resistant and antiviral drug therapies (notably, HIV and AIDS)

Generations of Peppered Moths Changed Color to Match Habitat

14 Origins of Life

NOTES
The ultimate spark of life may never be known, but science provides a controversial scenario of how life "might" have arisen.

Origins of the Universe & Earth

- First, the universe had to be formed — theoretically via the **Big Bang** about 16-18 billion years ago.
- Geologic and other physical evidence date the earth's origin to about **4.6 billion years ago.**
- The **crust** and **biosphere** (thin portion of earth where life exists) would not be habitable for nearly a billion years.
  - ◆ Up until then the earth was too hot to sustain life.

Earliest Life Forms

- How did the first cells form?
- Early hypotheses suggest life arose spontaneously from simple molecules (CO, CO_2, N_2, H_2O) that combined into larger, complex **macromolecules** such as proteins, carbohydrates, lipids and nucleic acids.

■ Some rocks from outer space (**meteorites**) have pre-formed complex organic molecules.

■ Whether life was seeded from outer space (**panspermia**), or macromolecules were synthesized entirely on earth, the next step was to incorporate these organics into **cells:** the **basic functional units of life.**

Primordial Soup

■ These first life forms were likely **heterotrophs,** which consumed the abundant food molecules present in the **"primordial soup"**

■ Later, photosynthesis (by **autotrophs**) developed and oxygen levels began increasing in the atmosphere.

■ The oldest fossils discovered (aged 3.8 billion years) consist of photosynthesizing bacteria called **stromatolites,** which still have representatives in colonies that form large, calcareous structures in some shallow, tropical oceans.

Stromatolites Form Aquatic Reefs

■ Oxygen Crisis & the Endosymbiotic Hypothesis

■ Geologic evidence supports increasing oxygen levels via photosynthesis created "rust" zones at similar ages in ancient sea beds worldwide.

◆ Chemically, oxygen is a corrosive element to organic molecules as well, and likely created a crisis for many of the earliest life forms.

◆ Some bacteria evolved a metabolic pathway that could neutralize as well as produce ATP energy from this highly—reactive oxygen.

◆ **Symbioses** formed between these oxygen-consuming, energy-producing bacteria and other larger, soft-bodied bacteria that lacked protection against the effects of oxygen.

■ This was the birth of the **eukaryotic** cell, from **prokaryotic ancestors;** one of the major evolutionary events in life.

■ This **endosymbiotic hypothesis** is supported by the following facts:

◆ **Mitochondria** (use oxygen for metabolism) have their own set of DNA, separate from that of the cell nucleus.

◆ **Mitochondrial DNA** is more like present-day bacterial DNA than the nuclear DNA of the cell in which it resides.

◆ **Chloroplasts** have their own genomes.

NOTES
Today, living organisms provide numerous examples of symbiotic relationships between single-celled organisms, sometimes including bacteria that perform the role of mitochondria in cells lacking ATP-producing organelles.

◆ Eukaryotic cells subsequently evolved into protists, fungi, plants and animals.
◆ Prokaryotes continued to thrive and, though microscopic, are among the most successful groups of organisms on earth.

Evolution of Eukaryotic Cells

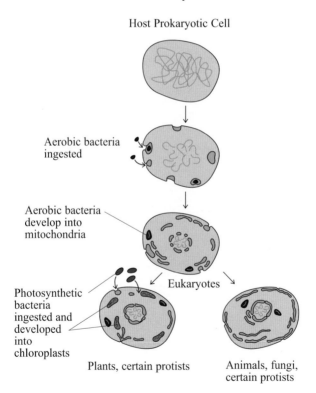

Host Prokaryotic Cell

Aerobic bacteria ingested

Aerobic bacteria develop into mitochondria

Eukaryotes

Photosynthetic bacteria ingested and developed into chloroplasts

Plants, certain protists

Animals, fungi, certain protists

Human Origins

NOTES

Where do humans fit in the evolutionary scheme? Some of the greatest evidence for evolution is seen when comparing **vertebrate chordates,** which include humans.

Comparative Anatomy of Adults

■ Obvious visual similarities in adult vertebrates link humans to other vertebrates, especially the great apes.

Examples:

- ◆ eyes
- ◆ ears
- ◆ mouth
- ◆ nose
- ◆ appendages

Comparative Embryology

■ **Earnst Haeckel** coined the phrase "ontogeny recapitulates phylogeny," suggesting the false claim that humans start as fish, then progress through a series of developmental stages that retrace the lower vertebrate groups before becoming human.

◆ Early developmental stages of humans share remarkably similar vertebrate characteristics that either disappear or become vestigial in adult humans.

• **Gill (pharyngeal) slits** (they occasionally do not close in infants—**cervical (branchial) fistulae**—may require surgery)

Embryonic Similarities Among Vertebrates

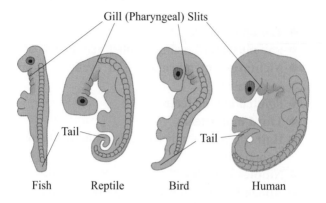

Fish   Reptile   Bird   Human

Vestigial Structures

■ Show clear links to vertebrate ancestry and include the following non-functional structures:

 ◆ **Tail bones (coccyx)**
 ◆ **Ear muscles** (function in other mammals)
 ◆ **Nictitating membrane** (third eyelid in some vertebrates)
 ◆ **Pointed canine teeth**
 ◆ **Third molar teeth**
 ◆ **Hair** (plays major thermoregulation role in most mammals)
 ◆ **Nipples in males**
 ◆ **Appendix** (functions as digestive caecum in many mammals)
 ◆ **Segmented muscles of abdomen**
 ◆ **Pyramidalis muscle**
   • Absent in 20 percent of humans
   • arguably unnecessary
   • present in other mammals

Some Vestigial Structures in Humans

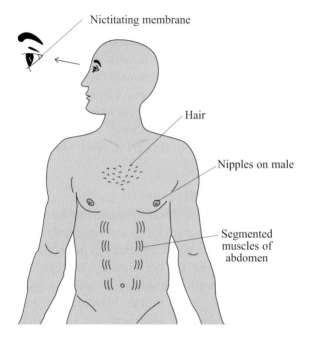

Molecular Comparisons

■ Comparison of DNA sequences in humans and chimpanzees show average similarity of 98.5 percent

■ Comparison of **hemoglobin/amino acid sequences** (the main carrier of oxygen in the blood of thousands of different animals (by itself evidence for evolution) between humans and other vertebrates show the same evolutionary patterns as those with skeletal/physical anatomy that is comparative, with the great apes showing the greatest similarity.

Fossil Record

■ **Fossils** show a transition from ape-like forms to the first primitive human forms that were truly **bipedal** (walking on the pelvic appendages or legs).

◆ Modern apes are not bipedal, but one of the oldest fossil forms (3.2 million years) resembling an ape to walk bipedally was named ***Australopithecus afarenesis*** or Lucy (named after a famous Beatles song).

◆ From this origin in Africa, modern humans, ***Homo sapiens,*** eventually arose.

◆ Debate exists among paleoanthropologists about how to arrange the phylogenetic tree of humans based on the available fossils.

◆ Most agree that Neanderthals were the most recent group of humans to become extinct, and were probably a subspecies called ***Homo sapiens neanderthalensis.***

◆ From these origins, humans have spread to most land areas on Earth.

Anthropoid Skeletal Comparison

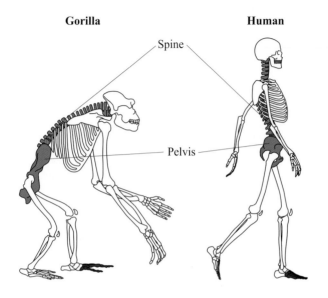

16 Molecular Biotechnology

NOTES
The discovery that DNA is the **informational** molecule housing genes started a revolution in biology. **Molecular biotechnology** is now a pervasive component in modern societies.

Cloning
Gene Cloning
- Making exact copies of genes
- Involves two major processes:
  - ◆ **Recombinant DNA**
    - Restriction enzymes create DNA fragments with the gene of interest.
    - DNA fragments are fused with DNA from a bacterium (**plasmid**).
    - Newly created recombinant DNA is placed into bacteria.
    - Bacteria produce protein for which the "cloned" gene is coded.
    - Large quantities of the gene, and thus protein, are produced as the bacterial cell reproduces.

Gene Cloning Using Recombinant DNA

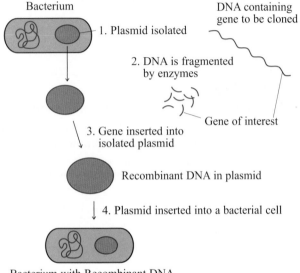

Bacterium

DNA containing
gene to be cloned

1. Plasmid isolated

2. DNA is fragmented
by enzymes

Gene of interest

3. Gene inserted into
isolated plasmid

Recombinant DNA in plasmid

4. Plasmid inserted into a bacterial cell

Bacterium with Recombinant DNA

◆ Polymerase Chain Reaction (PCR)

- **Amplifies** (copies) a segment of DNA without using a bacterial (or other) host organism.
- DNA sample is heated until the **double helix** denatures (hydrogen bonds are broken), separating the DNA into **two single strands.**
- Heat-resistant, single-stranded **DNA primers** allow DNA polymerase to add the appropriate nucleotides to each side of the separated DNA strands.
- This process results in **multiple copies** of the original DNA.
- Repeating the process on the copies, via automation, can amplify a small DNA fraction a billion fold in a short period of time.

Using PCR to Amplify DNA

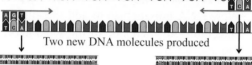

Separate DNA strands by heating

Primers add base pairs to DNA template strands

Two new DNA molecules produced

Repeat above processes to make multiple DNA copies

Reproductive Cloning

■ Produces **living cells/organisms** with **exactly the same DNA** in the nuclei as that from a donor cell/organism.

◆ Specifically, DNA from the nucleus of a **somatic cell** of the donor is inserted into an egg cell from which the original nucleus has been removed.

◆ The new **egg cell** is electrically or chemically stimulated to begin cell division and embryonic development.

◆ The growing embryo is implanted into a female where development continues until birth.

◆ The new individual is not a true clone of the donor organism, as the mitochondrial DNA is from the organism that donated the egg.

◆ Survival rates have been low as multiple factors (mostly unknown) influence successful development, such as source of donor cells.

Cloning

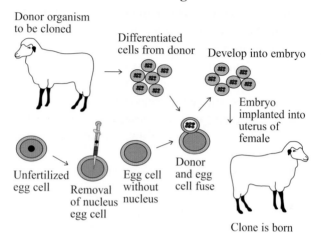

Donor organism
to be cloned

Differentiated
cells from donor

Develop into embryo

Embryo
implanted into
uterus of
female

Unfertilized
egg cell

Removal
of nucleus
egg cell

Egg cell
without
nucleus

Donor
and egg
cell fuse

Clone is born

■ Therapeutic Cloning

◆ Use of reproductive cloning to create human embryos to procure **stem cells,** which have potential to develop into adult tissues.

◆ These special cells may hold the key to treatments for many diseases (heart, cancers, Alzheimer's, Parkinson's) and afflictions (injury to spinal cord, including paralysis).

◆ Stem cells can also be retrieved from human embryos produced by regular fertilization processes (*in vivo* or *in vitro*) or adults (e.g., bone marrow).

◆ Stem cell procurement via cloning and embryos is a growing ethical and political issue.

Culturing Stem Cells

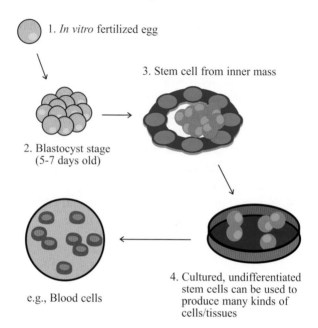

1. *In vitro* fertilized egg

3. Stem cell from inner mass

2. Blastocyst stage
(5-7 days old)

4. Cultured, undifferentiated
stem cells can be used to
produce many kinds of
cells/tissues

e.g., Blood cells

17 Genomics

NOTES
Genomics is the study of the structural and functional aspects of the entire set of genes in a species (i.e., **genome**)

■ Genomics encompasses many different aspects of approach:
  ◆ **Bioinformatics** uses computer/statistical applications to access large databases concerning DNA/gene/protein information
  ◆ **Proteomics** studies the functioning of the proteins coded by the genes
■ Several specific applications of genomics will be discussed further:
  ◆ **Restriction Fragment Length Polymorphisms (RFLP)**
    • Technique relies on enzymes discovered that protect bacteria from "foreign" DNA of **bacteriophages** (viruses specific for bacteria) and other invading bacteria.
    • These bacterial restriction enzymes cut foreign DNA at specific points or **restriction sites,** while protecting their own DNA by adding special "buffering" functional groups to potentially susceptible areas.

- Exact positions of restriction points are highly individual, reproducible and measurable.
- DNA samples from the same individual will produce the same fragments, but these fragments will be different from others **(polymorphic).**
- Fragment patterns can be represented visually as a **DNA fingerprint,** by use of special electrophoretic processes.
- RFLP is used frequently in forensic, criminal and paternity applications.
- Because DNA samples may be minute in some of these applications, PCR amplification may be used to create quantities necessary for RFLP analysis.
- A modified DNA fingerprint approach has been developed using polymorphisms of **satellite (repetitive) DNA regions** called **Simple Tandem Repeats (STR).**

DNA Fingerprinting Using RFLP

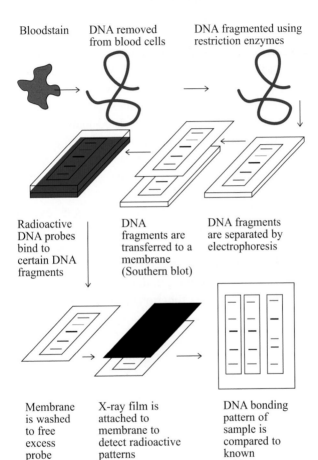

Bloodstain

DNA removed from blood cells

DNA fragmented using restriction enzymes

Radioactive DNA probes bind to certain DNA fragments

DNA fragments are transferred to a membrane (Southern blot)

DNA fragments are separated by electrophoresis

Membrane is washed to free excess probe

X-ray film is attached to membrane to detect radioactive patterns

DNA bonding pattern of sample is compared to known subjects

■ Human Genome Project

◆ Monumental, historical effort to determine the actual sequence of the entire set of chromosomes in humans: **gene mapping**.

◆ Involved **over 3 billion base pairs,** which if written, would create a book with a half-billion pages and take nearly a lifetime to read.

◆ Several molecular techniques were employed, with automated computer-assisted analysis paving the way for a rapid conclusion to the project.

◆ Although the precise number of genes is still unknown, *a priori* estimates suggested there would be nearly 100,000.

◆ Actual number probably does not exceed 40,000, which when compared to simpler organisms suggests human genomics is extremely concise, but complex.

◆ Future studies will undoubtedly reveal much about how genes function, which should lead to numerous future benefits.

Genomic Project – Mapped Human Genes

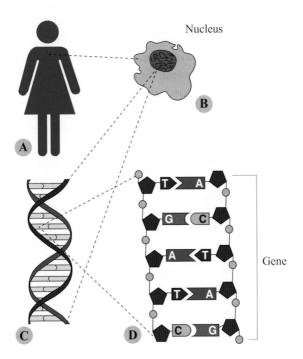

■ Gene Therapy

- ◆ Treating diseases and injury in humans involves the use of harmless retrovirus vectors (or other entry mechanisms) that possess the enzyme **reverse transcriptase,** allowing them to insert genetic information "into" DNA.
- ◆ Normal information flow occurs "from" the DNA.
- ◆ These treatments raise ethical questions, but certainly have tremendous potential.
- ◆ Limited success and legal restrictions using human subjects have made progress in this area challenging.

Using Retroviruses to Insert Healthy Genes

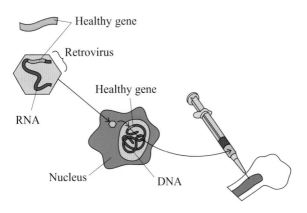

◆ **Genetic Engineering**

◆ Research involving gene transfer in non-human organisms has been much more extensive.

◆ **Transgenic** and **genetically modified** plants and animals are becoming more common.

◆ Great potential to artificially select desirable traits in crops, farm animals, etc.

◆ Safety concerns are still high as this new technology is incorporated into modern society.

18 Biology of Cancer

NOTES

◆ Cells reproduce by dividing primarily through two processes:
  - **Mitosis:** Nuclear division
  - **Cytokinesis:** Cytoplasm division [*see* chapter 6, **Cell Reproduction**, for review]

▓ Cell division is part of the **cell cycle,** which is under a control system involving internal and external factors.

Cancer Cells: Transformation by Uncontrolled Cell Division

▓ **Cancer cells** have escaped this regulatory process through **transformation** and divide uncontrollably.

Tumors: Benign to Malignant

▓ This causes **tumors** to form, which may progress from a **benign** to a **malignant** state and interfere with normal tissue functioning.

Tumor Formation & Spreading

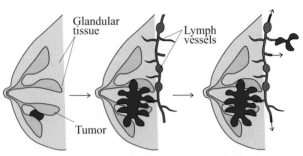

1. Malignant tumor starts from single cancerous cell

2. Tumor grows, invading neighboring tissue

3. Lymph and blood vessels spread cancer cells to other areas of the body

■ **Metastasis:** Initial tumor cells can spread and form more malignant tumors in other tissues in the body.

■ **Oncogenes:** Stimulate abnormal cell growth and division, which can lead to malignant tumors.
  ◆ These abnormal genes are converted from normal genes **(proto-oncogenes)** that regulate the cell cycle.
    • Viruses can also deliver oncogenes to cells.

Oncogene Activation Leading to Cancer

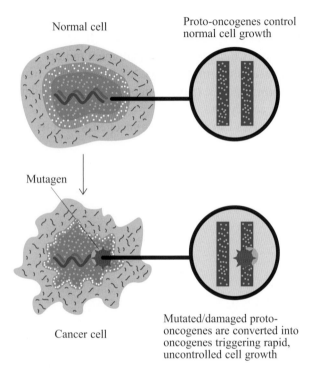

Normal cell

Proto-oncogenes control normal cell growth

Mutagen

Cancer cell

Mutated/damaged proto-oncogenes are converted into oncogenes triggering rapid, uncontrolled cell growth

Tumor-Suppressor Gene Deactivation
Leading to Cancer

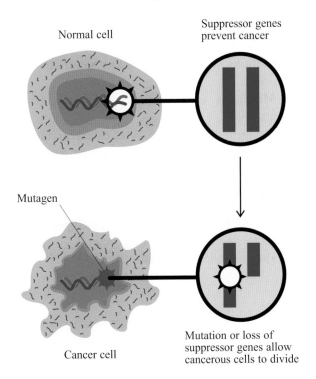

Normal cell

Suppressor genes
prevent cancer

Mutagen

Cancer cell

Mutation or loss of
suppressor genes allow
cancerous cells to divide

■ **Tumor suppressor** genes normally prevent the uncontrolled growth and division of cells and tissues.

Mutations

■ **Mutations** are primary factors contributing to cancers.
  ◆ **Mutagens** are any factors that can trigger mutations.
  ◆ Mutagens that cause cancer are called **carcinogens.**
■ All tissues in the human body are susceptible to tumors because mutations (either induced by carcinogens or **inherited**) can occur in any cell.
■ Cancers are prevalent and difficult to cure (in most cases) because of our limited knowledge about:
  ◆ Factors controlling the cell cycle
  ◆ The genomics of humans

Biology of Aging

NOTES

Most animals in nature die shortly after their reproductive years, and in some cases, die immediately after reproduction.

Human Lifespan

■ Humans and most animals kept under controlled conditions can survive many years after fertility has waned, allowing the phenomenon of aging to be studied.

♦ For humans, the potential to live longer has been realized during our history. In the last 50 years, the average lifespan in well-developed countries has risen from the 60s-70s to nearly 80 years of age.

■ Considering the longevity of some rare individuals, human lifespan could be 120 to 130 years in the near future.

Human Lifespan Increase

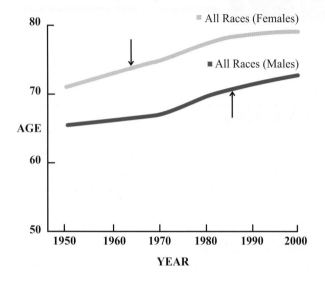

Theories of Aging

NOTES

Keep in mind some questions pertaining to aging:

◆ What prevents all but a few of us from living to our physiological maximum?

◆ What are the specific causes for the physical transformations that occur as we age?

Random Events

■ **Random events** may accumulate and contribute to early **senescence;** some specific hypotheses follow:

◆ **Free radical** formation typically involves the production of oxidative metabolic by-products such as molecular variants of oxygen, which may damage the DNA, RNA, proteins and mitochondria.

• Antioxidants produced naturally may eventually lose the battle in cells, causing cell death.

• Proponents of this hypothesis suggest supplemental intake of antioxidants (e.g., found in vitamins) may slow this form of damage.

◆ **Cross-linking** suggests that as cells age, structural molecules such as DNA and proteins form unsuitable attachments within or between other molecules.

- Skin wrinkling, cataracts of the eye, atherosclerosis in blood vessels, kidney function and brain function decline are all possibly related to cross-linking.
- Some drugs that prevent or slow cross-linking may be important future therapies.

◆ **Wear and tear** suggests that the mere use of cells and concomitant damage result in aging:

- This type of damage occurs at the DNA level, which has its own set of repair proteins.
- Years of exposure to mutagens such as toxins and various forms of radiation are not always repaired.
- At the ends of DNA molecules are protective caps called **telomeres,** which are degraded with each cell division event.
- Telomere loss eventually can lead to DNA damage.
- Telomerase, an enzyme that repairs these end caps, has been shown to keep cells in a more "youthful" state.

◆ **Somatic mutations,** those occurring in tissues outside of the egg or sperm, could lead to diminished function; skin and connective tissues lose resiliency, muscles become weaker, brain cells become less efficient, etc.

◆ **Rate of living hypothesis:** Suggests those that "live the fastest, die the youngest."

Examples:

Rate of Living Hypothesis

• Theorizes those organisms with the most active metabolisms have the shortest lifespan.

• With mammals, this is usually the case (e.g., an elephant lives longer than a mouse).

• Hypothesis may be broadly linked to those under the pre-programmed events [*see* page 109].

Physical Changes During Aging

30 years   40 years   60 years   80 years

■ **Pre-programmed events** may be a cause of senescence in humans; following is a discussion of specific hypotheses:

◆ **Genetic theory** suggests our lifespan is determined by the inherited genes

- When food and health issues are maintained at least minimally, humans have roughly the same lifespan.
- Females in most instances (including other animals) typically live longer than males.
- Offspring of long-lived parents typically live longer than offspring of shorter-lived parents.
- The above observations strongly suggest at least part of lifespan determination is related to **longevity assurance genes.**

◆ **Pacemaker theory** suggests there are **"biological clocks"**—or **pacemakers**—that commence at birth and simply slow and stop, ending in death.

- Specifically, the immune and neuroendocrine systems are thought to be controlled by pacemakers.
- Cessation of these systems could account for body-wide failures, susceptibility to attack by foreign agents, and increase incidence of cancers.

Immunology

21

NOTES
The body has two main lines of defense
against injury and infection: nonspecific and
specific immunity.

Nonspecific & Specific Immunity

◆ **Nonspecific immunity** involves a generalized,
similar response to a wide variety of potentially
harmful conditions. A typical component of this
response is **inflammation,** which results in
swelling, redness, heat and pain in the affected area.

◆ Specific immunity is an extremely specific
response typically involving the production of
antibodies, which are designed with the exact
purpose of combining with specific cell surface
markers, or **antigens,** of foreign agents
(microbes, toxins).

Active & Passive Immunity

■ Selected subjects related to immunity are discussed as follows

◆ **Passive immunity** involves receiving antibodies or antiserum from another source.

- This could involve maternal antibody delivery to the fetus/child via breast milk from the mother or injections (also for treatment of venomous bites/stings).

Antibodies Injected or Passed to Others

Active Immunity   **Passive Immunity**

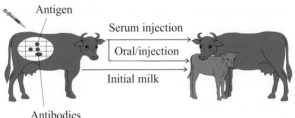

Antigen

Serum injection

Oral/injection

Initial milk

Antibodies

◆ **Vaccinations** contain weakened versions of pathogens injected into the body to stimulate, among other aspects of specific immunity, B cells to produce two products:

- **Plasma cells,** which begin synthesizing antibodies within 10 to 17 days.
- **Memory cells,** retain the potential (for up to many years) to develop quickly (within 2 to 5 days) into antibody-producing plasma cells upon subsequent exposure.
- This quicker response could mean the difference between successfully destroying the foreign antigen versus possible death of the individual.

Allergic Reaction Events

1. Immune system exposed to pollen through nose, lung or eyes

2. Antibodies specific to the pollen grains are formed

3. Antibodies bind to mast cells in connective tissues

5. The released chemicals trigger allergic reaction, (runny nose and eyes, itching throat and nose, sneezing, respiratory congestion, related asthma symptoms)

4. Pollen again enters the body, attaching to antibodies, triggering the mast cells to release histamine and other chemicals

◆ **Allergies** are hypersensitive tissue reactions to part of the specific immune response.
  • Specifically, antibodies against specific antigens called **allergens** trigger tissue response resulting in typical allergic symptoms (e.g., hay fever, asthma).
  • Severe allergic reactions can lead to **anaphylactic shock,** which may be life threatening.

Autoimmunity

◆ **Autoimmunity** is a condition in which cells of the specific immune response attack healthy tissues.
  • Normally, those antibodies and cells of the immune response that could harm "self" tissues are either suppressed or deleted to prevent such self attacks.

  • The following diseases/afflictions are triggered or related to autoimmunity:
    > Rheumatoid arthritis
    > Diabetes mellitus
    > Grave's disease
    > Multiple sclerosis
    > Lupus

Autoimmune Disease Leading
to Rheumatoid Arthritis

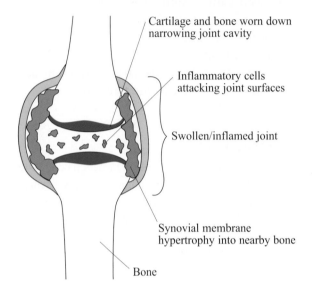

Cartilage and bone worn down
narrowing joint cavity

Inflammatory cells
attacking joint surfaces

Swollen/inflamed joint

Synovial membrane
hypertrophy into nearby bone

Bone

◆ **Immunodeficiency** diseases are those in which some aspect of the immune system (usually specific) is defective, thus compromising the ability of the body to protect itself.

  • One of the best known of these is **Acquired Immunodeficiency Syndrome (AIDS)**—a disease which is triggered by the **Human Immunodeficiency Virus (HIV).**

  • In this affliction the virus attacks immune cells called helper T cells, which are integral in mounting a specific immune response

  • Individuals with such compromised immune systems are susceptible to secondary infections and cancers, which untreated usually leads to death.

  • AIDS is still a worldwide health issue and the leading cause of premature death in some countries.

◆ **Severe Combined Immunodeficiency Syndrome (SCIDS)** is a rare congenital condition in which T and B cells are defective.

  • In the most severe cases, a person is born essentially with no specific immune response and stands little chance of warding off infection.

  • Death can occur within the first year without a bone marrow or stem-cell transplant

HIV Virus Attacking Helper T-Cell Lymphocyte

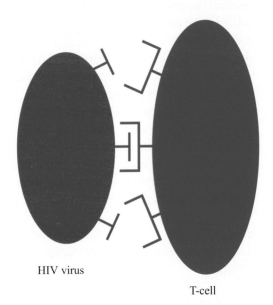

HIV virus

T-cell

◆ **Bacterial resistance to antibiotics** can occur when medical drugs are used to supplement the specific immune response, the latter of which may be too slow to prevent serious and possibly fatal symptoms.

• When antibiotics are taken, highly resistant forms of bacteria may survive and reproduce.

• These new "resistant" strains may be extremely difficult, if not impossible, to treat.

• **Over-prescribing** of antibiotics may be a leading cause of resistance.

• As much as half of the roughly **100 million** prescriptions for antibiotics written each year may be unnecessary.

Examples:
Cold and flu symptoms are caused by viral infections; therefore, antibiotics are of limited use in fighting them.

• When prescriptions are given, medication should be taken to completion—taking only a portion of the pills may allow the hardiest bacteria to survive and evolve.